WORLD OF DISCOVERY

CLEVER CAMOUFLAGERS

by Anthony D. Fredericks

NorthWord

NorthWord PRESS, INC.
Minocqua, Wisconsin

DEDICATION

To George DeSau—

A very clever un-camouflager!

ACKNOWLEDGMENTS

A special note of thanks

to all my students—

who continue to examine,

invent, and promote

all the creative possibilities

of non-fiction literature.

Photography © 1997: Art Wolfe, Front Cover, 15, 16; Jim Brandenburg/Minden Pictures, 3; David Hosking/Dembinsky Photo Associates, 4; E. R. Degginger/Bruce Coleman, Inc., 5, 8; Martin Withers/Dembinsky Photo Associates, 7; Jeff Foott/Bruce Coleman, Inc., 10; Mark J. Thomas/Dembinsky Photo Associates, 13; Norbert Wu/Mo Yung Productions, 18; Frans Lanting/Minden Pictures, 21; Dan Dempster/Dembinsky Photo Associates, 23; Alan G. Nelson/Dembinsky Photo Associates, 25; Peter Parks/Mo Yung Productions, 27; Jim Manhart, 29; Dean Lee/The Wildlife Collection, 31.

NorthWord Press, Inc.
P.O. Box 1360
Minocqua, WI 54548

Illustrations by Kay Povelite
Book design by Kenneth Hey

For a free catalog describing our audio products, nature books and calendars, call **1-800-356-4465**, or write Consumer Inquiries, NorthWord Press, Inc., P.O. Box 1360, Minocqua, Wisconsin 54548.

Library of Congress Cataloging-in-Publication Data
Fredericks, Anthony D.
     Clever Camouflagers / by Anthony D. Fredericks.
       p.   cm. — (World of discovery)
      Summary: Tells how twelve animals from around the world change their appearance to avoid predators.
     ISBN 1-55971-585-5 (softcover)
      1. Camouflage (Biology)—Juvenile literature.   [1. Camouflage (Biology) 2. Animal defenses.] I. Title. II. Series.
QP767.F73 1997
591.47'2—dc21
                            96-37142

Printed in Malaysia

CLEVER CAMOUFLAGERS

CONTENTS

Walking Stick

About Camouflagers

Chameleon

Have you ever played a game of "hide and seek"? Did you try to hide from the person who was "it" by standing behind a tree or scrambling underneath something? If you were "it," did some of your friends make noises or have brightly colored clothing that made it easy for you to find them?

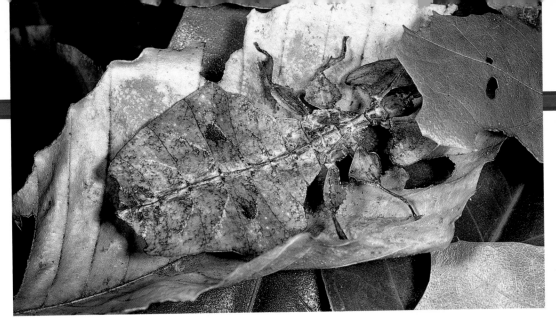

Leaf Insect

Many animals play "hide and seek" too, but for very different reasons. Some animals hide themselves to avoid being eaten by their enemies. Other animals disguise themselves so they can sneak up on their food and capture it. Scientists refer to this hiding ability as **camouflage**.

Several kinds of animals have colors or shapes that help them look like something else in their environment. There are fish, frogs, and insects that look just like leaves; grasshoppers that look like rocks or stones; moths and birds that resemble pieces of wood; lizards that look like trees or plants; and caterpillars that look like bird droppings.

Animals camouflage themselves for two reasons: to find food or to avoid becoming food for something else.

Scientists tell us that the colors and shapes of animals have come about (or evolved) over thousands or millions of years. An animal is able to survive because it has adapted to its environment—it knows how to locate its food and what to do to hide from its enemies. To do that some animals have developed distinctive means of camouflage—pretending to be something they are not.

In this book you will discover animals that play "hide and seek" every day of their lives. You'll learn about a bird that is white in winter and brown in summer, a frog that looks just like a leaf, a fish that looks like seaweed, and a lizard that changes its color a dozen times a day. For you, "hide and seek" may be a game; but for many animals, it is the only way they can survive.

Colorful Creatures

Stick out your tongue. How long is it? Like most people your tongue is probably 3 to 4 inches long. And, like most people, you use your tongue to help you eat and taste your food. What if your tongue were as long as your whole body and you could use it to capture your food and bring it into your mouth? That would be an incredible ability!

One amazing animal—the chameleon—not only has that wonderful skill, but also has another talent—one for which it is better known. In fact, when most people think about animals that are able to camouflage themselves, chameleons often come to mind.

As you may know, chameleons can change their body color to match their surroundings. For example, if a chameleon is resting on a green plant its skin color becomes green. If it is resting on a brown tree, its body color changes to brown.

A chameleon can change its color to shades of red or black as well.

Chameleons also change their skin color in response to the intensity of sunlight or the temperature of the air, or even their emotions. Not only do chameleons change their colors to blend in with their surroundings, they also do so to let other chameleons know how they feel.

Depending on the species, chameleons range in size from 2 inches to 2 feet long.

A chameleon's body is shaped like a leaf and ends in a prehensile (able to grasp objects) tail often held in a tight coil. Chameleons are very slow climbers, taking their time in moving through their environment.

One of the most distinctive features of a chameleon is its eyes—each of which can move independently from the other. Swiveling its eyes up and down lets the chameleon observe two separate objects at the same time or look at the same target from different angles. This ability is particularly handy when the chameleon looks for food.

Chameleons typically prey on insects, spiders, scorpions, and other small invertebrates (animals without backbones). Some of the larger species of

If a chameleon loses a fight with another chameleon its skin turns dark green. If it is angry, its skin color changes to black.

chameleons, however, will attack small birds, mammals, and lizards. The chameleon is very patient. When it locates a potential meal it will look at it for some time before attacking.

Powerful muscles and a special bone in the chameleon's mouth make the attack quick. Its tongue shoots out, trapping the victim on the tip, and carries it back into the chameleon's mouth.

While this animal is one of nature's most unusual, it is also one in danger. Because it lives in the rain forests of the world, its habitat is being destroyed at an alarming rate. As the rain forests are being reduced, so is the chameleon's home. Saving the rain forests will help ensure its survival.

Leaf insects belong to a distinctive family of insects called Phasmids. Most of the members of this group of creatures inhabit forests and woodlands throughout the world—mostly in tropical regions. Many species can be found in the wooded areas of Australia and Southeast Asia.

Plant Pretenders

How would you like to be invisible, or almost invisible? How would you like to be able to sit in a classroom without being seen or walk through a shopping mall and not be noticed?

There is an insect that is practically invisible simply because it looks like a leaf. This extraordinary creature, the leaf insect, copies vegetation so closely that it is almost impossible to see, even when viewed close up. Since leaves are everywhere in the insect world, this characteristic has helped ensure the survival of this unique animal.

There are about 2,000 different species of leaf insects found throughout the world. They range in size from less than 1 inch to 1 foot long.

The eggs of leaf insects are hard-shelled and often look exactly like the seeds of the plants on which they feed. A female leaf insect will often lay hundreds of eggs a day. These eggs need four to six months before they hatch (depending on the temperature).

Young leaf insects are often marked and colored to look like glossy, green leaves. Their bodies are covered with branched vein patterns similar to those on a living leaf.

Occasionally, these animals have two or three brown spots on their bodies resembling the marks left on leaves by disease or the

Fantastic Fact

Some species of leaf insects can change their color throughout the day—becoming light in the daytime and dark at night.

nibbling of other insects.

In fact, the appearance of young leaf insects matches their surroundings so closely that other animals can crawl right over them and not even know they are there.

Older and larger leaf insects also use color to camouflage themselves. As they mature, leaf insects turn brown, tan, or speckled in color—closely matching the colors of dead leaves scattered on the forest floor.

The limbs of these amazing creatures also resemble leaf parts. The legs and head are often shaped like flat extensions of a plant leaf or may even resemble the partially

eaten parts of tree foliage (leaves).

During the day, leaf insects remain perfectly still for long periods of time. At night, however, they become active—seeking leaves and other vegetation. If startled by an approaching predator, a leaf insect can remain motionless for several hours looking just like every other leaf on the ground.

Flattened bodies, wings and legs with scalloped edges, green and brown coloring, and very slow movements allow this "plant pretender" to survive and thrive in a very hostile environment.

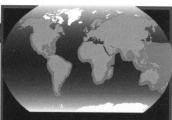

There are more than 150 different species of pipefish throughout the world. They can be found in the shallow waters of tropical and temperate seas, although a few live in depths of 50 feet or more.

Very Vertical Varieties

How long can you stand on your head? One minute? Two minutes? Five minutes? How do you think it would feel if you had to spend almost your entire life standing on your head? Well, you're about to meet a most unusual creature that does just that.

The pipefish is a slender fish with a long head and a tubular ("0"-shaped) mouth. Different species of pipefish range in size from 1 inch to 18 inches long.

**Unlike most fish,
pipefish can move
their eyes independently
of each other.**

Instead of scales, pipefish have a series of jointed bone-like rings encircling their bodies from their heads all the way down to the tips of their tails. These rings help pipefish maintain a pencil-like shape. Depending on the species, a pipefish's colors may range from bright green to drab olive.

Pipefish spend almost their entire lives swimming in a **vertical** (up and down) position. With their heads pointed toward the bottom of the ocean they look almost exactly like seaweed. Not only does this physical feature camouflage them from their enemies, it also allows them to

sneak up on their prey. Occasionally, however, pipefish do swim horizontally, especially if they need to escape an enemy.

Pipefish are related to sea horses; and they raise their young in similar ways. For example, some pipefish males have small pouches under their bellies in which the females lay eggs. In other species, the females lay their eggs on the underside of the males where they stick until hatching. Just like sea horses, the males typically take care of the young fish after they hatch.

Pipefish have no teeth, nor do they have a true mouth. Their jaws are locked into place—forming a permanent and unmovable opening. They eat by sucking small animals such as plankton directly into their stomachs.

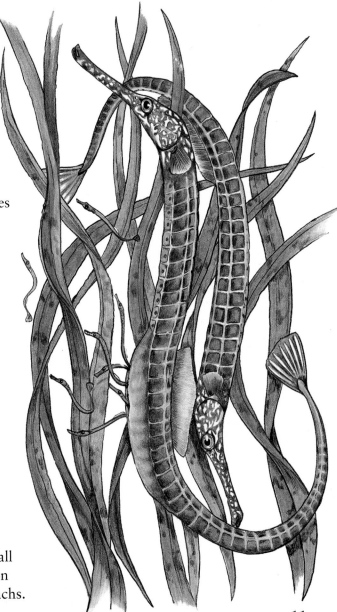

Servals can be found throughout Africa—particularly south of the Sahara Desert. They prefer to live in well-watered and open grassland areas of the continent.

Careful Creatures

Many animals have permanent markings, colors, or shapes that help them look like parts of the environment in which they live—a benefit in avoiding detection by their enemies. These animals take advantage of a biological feature known as **protective coloration** (skin or fur colorings which allow them to fade into the background and hide). This is particularly useful for young animals that are not strong enough or fast enough to escape their enemies.

One animal that exhibits this remarkable trait is the serval, a member of the cat family. At first glance, this animal appears to be a miniature cheetah. It has long legs, just like a cheetah, which help it move quickly through tall grass and swampy land. Its very large ears, which can turn in many different directions, help it locate the sounds of nearby animals. These features help it capture its prey, such as small mammals, rodents, birds, and lizards.

A serval grows to be 1 1/2 feet in height and up to 5 feet long. It weighs about 20 to 40 pounds when fully grown. But it is its coloration that is most distinctive—particularly the colors of the young.

The short hair of a young serval is yellowish-brown on its sides and back—often with a pattern of black stripes and spots. These colors are so similar to the colors of the short grassland areas in which it lives that young servals, which can remain motionless for long periods of time, are very difficult to detect. This protective coloration allows the parents to go off hunting while the babies remain behind—carefully and cleverly camouflaged in a nest of grass. Potential predators often pass by without noticing the young ones at all.

Young animals of many species are particularly defenseless for the first few days or weeks after birth. Young servals have been able to survive this difficult time in their lives by taking advantage of a unique form of camouflage. By hiding from their enemies they increase their chances for survival in the wild.

Servals enjoy water and will often swim after frogs and fish in small pools or marshes.

The casque-headed frog is found primarily in the South American country of Ecuador. Other varieties of these frogs are located in the jungles of Southeast Asia and the rain forests of South America. They average about 2 to 3 inches long.

Hideous Hiders

One look at this creature and you might think you were observing an alien from another planet or some monster in a science-fiction movie. However, the strange appearance and unusual features of this animal help it survive from day to day.

The casque-headed frog is one of a group of several frogs whose shape, color, and appearance make them look exactly like dead and fallen leaves in the jungles where they live. Extra flaps and folds of skin jutting out in several directions, a pointed snout, brown and mottled skin,

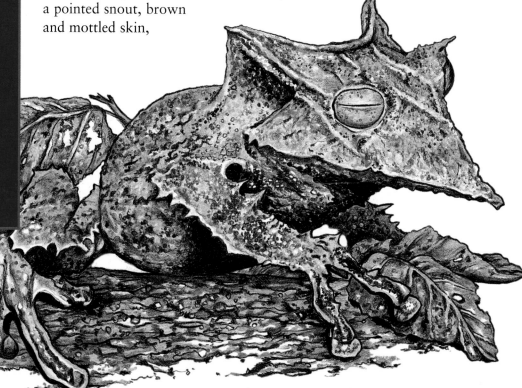

either making its home among the leaf litter that covers the forest floor or living up in the branches of rain forest trees. There it can rest or remain motionless for long periods of time—blending into the natural background.

The casque-headed frog also has another unique trait. After mating, the female frog carries all her eggs on her back for many days. There, the eggs hatch directly into young frogs. The tiny frogs stay "on board" their mother until they are old enough to venture out on their own.

and a bumpy body help this creature blend in with the plant life in which it lives.

As a result, this curious frog appears almost invisible among leaves on the ground.

Its camouflage helps the casque-headed frog in two ways. First, the frog easily escapes detection from any potential enemies. Snakes and other tropical predators like nothing better than a tasty frog for dinner, and this frog's camouflage helps it avoid

becoming part of a snake's diet. Second, their camouflage also allows them to wait, without being seen, for their own food to arrive.

Like most frogs, the casque-headed frog is insectivorous (they primarily eat insects). Crickets, ants, or termites that get too close to this well disguised creature quickly become its next meal.

The casque-headed frog is primarily a terrestrial animal. That means that it spends its entire life on land rather than in the water—

All frogs must close their eyes in order to swallow. They use their eyeballs to help push food down their throats.

The orchid praying mantis inhabits the jungles of Southeast Asia. It is found mostly on the island nation of Malaysia. Its existence, however, is threatened due to the destruction and deforestation of large tracts of rain forest land in that tropical country.

Ferocious Flowers

If you were to go into your garden or the garden of a neighbor, you might see a wide variety of pretty flowers. Many of those flowers would have beautiful smells, and lovely blossoms. But if you were in Southeast Asia, you might want to think twice about bending down and smelling the blossoms of a wild rhododendron—particularly if you were a small (and tasty) insect.

The orchid praying mantis lives among the blossoms of the wild rhododendron bushes typically found in this tropical environment. Because of its color, shape, size, and special features this insect is almost impossible to distinguish from the bunches of flowers on this plant.

The mantis' forelimbs closely resemble parts of one of the flowers in a group of closely packed blossoms. Its body parts look

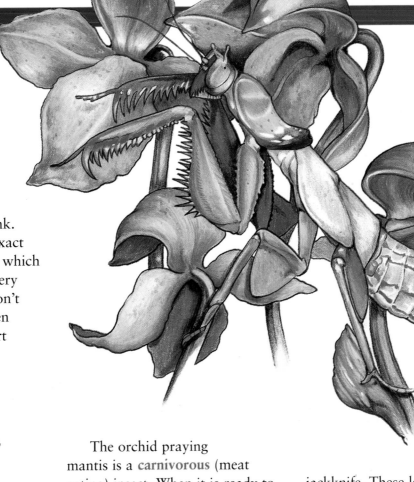

exactly like a nearby blossom. And its hind legs closely match the parts of yet another flower in the bunch.

The orchid praying mantis is colored a delicate shell-pink all over—even its eyes and antennae are pink. As a result, it takes on the exact coloration of the flowers on which it rests. Unless you looked very closely (something insects don't often do) you would not even know this animal wasn't part of a flower.

All praying mantises eat their victims alive. And females sometimes eat their mates.

The orchid praying mantis is a **carnivorous** (meat eating) insect. When it is ready to eat, it waits patiently on the rhododendron flowers. Flying insects looking for nectar approach the flowers and are quickly seized with a lightning-quick movement by a pair of powerful pink claws. The long front legs of a mantis can close rapidly at the joints, snapping shut like a jackknife. These legs also are armed with several rows of teeth. Once a victim is caught by a praying mantis, it cannot escape.

This insect looks so much like a flower that it is able to stay in the same place for days and capture all the food it wants. Its dinner is always delivered by air—right to its table.

The sea dragon is a species of sea horse that lives in and around the coral reefs of Australia. Other sea horses live in shallow coastal areas throughout the world.

They range in size from the dwarf sea horse (1/2 inch long) to the Pacific sea horse (1 foot long).

Delicate Dragons

Living in the ocean can be very dangerous. Large fish and other marine creatures are always on the lookout for smaller organisms to turn into a tasty meal or a quick snack. Some sea animals can swim rapidly from their enemies. Others can dart in and out of rocks and plants. But one of the most unusual "hiders" in the sea may be the sea dragon, a species of sea horse that inhabits tropical oceans.

One of the most distinctive features of a sea horse is that it has a hard bony skeleton on the outside of its body as well as another skeleton on the inside. Its shape makes it look like a horse swimming through the water. Its scientific name—*Hippocampus*—means "horse-caterpillar."

The sea dragon has a long spiny body, bony head, and tubular snout. But it also has something else—lots of

A sea dragon moves
the fan-shaped fins on
its back at a speed of
2,100 beats per minute.

seaweed-like growths covering its
entire body. It even has growths
coming out of its nose, its head,
and along the sides of its body.
Unless you looked very carefully,
you would think that it was just
another piece of seaweed floating
in the water.

Because the sea dragon looks so
much like seaweed it is able to hide
from enemies such as the sea turtle.
It is able to curl its prehensile tail
around a plant and remain perfectly
still for long periods of time, sway-
ing and moving with the plants so
that it is barely recognizable.

Sea dragons travel by
filling an internal swim bladder
with air (so they rise in the water)
or by emptying the bladder (so they
sink). By changing the amount of
air in their swim bladders, sea
dragons can go up and down at
will. To swim forward or back-
ward, they use the fan shaped fins
on the back and sides of their thin
bodies. Even when sea dragons
swim they are difficult to see,
because they travel very slowly,

and only their **transparent** (see-
through) fins move.

The sea dragon has eyes that
work independently of each other.
That means one eye can be looking downward while the other eye
is looking upward. This is useful
in locating food or any potential
predators.

Leaping Lizards

If you were to walk through a tropical rain forest you might notice that the trees have lots of bumps and ridges. Many of these occur naturally on the bark of a tree, but some may not be related to the trees at all. In fact, they are often separate organisms such as plants or animals. One of the most unusual animals to make its home on the trunks of rain forest trees is probably one you would scarcely notice—that is, until it "flew" through the air to another tree.

This distinctive animal is the flying gecko, a creature whose coloration makes it look exactly like the bark of the trees on which it lives. The flying gecko's colors naturally blend in with the brown and black **mottling** (spotted coloring) of those trees. Even its skin has the patterns and indentations of tree bark.

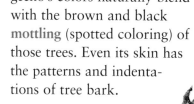

The flying gecko can remain motionless and unseen until a tasty insect comes too close and is quickly gobbled up. Geckos have very large eyes and well-developed hearing organs which assist them in locating their food, particularly at night when they are most active.

Although flying geckos are cleverly camouflaged against tree surfaces, they have also developed a distinctive way to **elude** (get away from) any would-be enemies. Besides a broad leaf-like tail, these creatures have wide flaps of skin along the sides of their bodies and narrow skin flaps along the

sides of their heads. When threatened, the flying gecko leaps into the air and spreads its legs out to the sides so that they act like the wings on a glider. By making its body flatter and wider it falls through the air like a parachute. In this way it can glide from tree to tree and easily escape.

Another interesting feature of this animal is its ability to cling to almost any surface. The flying gecko's feet have broad toes covered with ridges of scales. These scales have thousands of microscopic hooks that can hang onto almost any surface. As a result, the flying gecko can run over, under, and around its environment to chase after a meal or to escape another animal.

Its special feet even allow it to hang upside down from tree

branches or smooth surfaces for long periods of time.

Geckos range in size from 1/2 inch to 14 inches. These remarkable creatures also have the ability to break off their tails when attacked. A new tail grows in its place in a few months.

A flying gecko is able to lick and clean its eyes with its tongue.

Stalking Sticks

Walking sticks are one of about 2,000 species of stick insects that live throughout the world—many of them in tropical regions.

STOP! FREEZE! Sit perfectly still! Don't move a muscle! How long do you think you can hold that position? If you're like most people, you find it very difficult to stay motionless for any great length of time. But there is an amazing creature that probably lives in your back yard or in a nearby forest that can remain motionless for many hours. Not only can it stay perfectly still, but its body shape makes it look exactly like the twig or branch on which it rests. As a result, this creature—the walking stick—is almost invisible.

Walking sticks are members of a family of insects with the scientific name *Phasmidae*. This is a Greek word which means **apparition** (ghost or phantom).

Walking sticks are extremely thin, with long spindly legs and compressed bodies. Depending on the species, they grow to be 2 1/2 to 4 inches long. Their antennae may be 1 3/4 to 2 1/2 inches long. Typically

Fantastic Fact

Some species of walking sticks reproduce by parthenogenesis (the ability of a female to lay fertile eggs without mating). There is a species in New Zealand in which males have never been discovered!

they are green or brown so that they look exactly like a twig, stick, or plant stem. In fact, all their body parts resemble plant parts. When they walk, these curious creatures look like moving twigs. When they stand still, which they do quite often, they look like an extension of a branch or stick.

Many species of walking sticks can change their colors according to the season of the year. In spring or summer they are light to dark green, and can easily hide among the leaves of trees. In fall and winter they are able to change their colors to shades of brown and gray, concealing themselves among the branches of deciduous trees (those that lose their leaves in autumn).

The walking stick is quite common throughout the United States. Although it is hard to see in its natural environment, its eating habits can easily be detected. In fact, it is considered a pest in some sections of the country simply because it defoliates (eats the leaves from) large tracts of trees.

Although the walking stick is cleverly camouflaged in its natural environment, it is often located by predators such as birds, midges (small blood-sucking insects), and wasps. If captured, it has a unique trick to escape: It can break off one of its own legs—and leave the attacker with a very small body part to eat. The walking stick later grows a new leg.

PTARMIGANS

Ptarmigans can be found in all the higher latitudes of the world including Scandinavia, Siberia, Alaska, northern Canada, Greenland, and Iceland.

Feathered Fashions

How would you like to be able to naturally change the color of your hair whenever you wanted? What if you could change your hair to blond or white in the winter, to black or brunette in the spring and summer, and to gray in the fall? That ability would be most unusual, especially if you could do it year after year. Well, you're about to meet an animal that does that—the ptarmigan.

The ptarmigan is a medium-sized bird that lives in some of the coldest regions of the Northern Hemisphere. This environment is often barren—having nothing more than large open expanses of rocks, short vegetation, and long rolling meadows. The climate in these regions is particularly harsh, with temperatures constantly below freezing. Snow and ice cover much of this land for most of the year.

Because of this open environment and lack of good hiding places, animals such as the ptarmigan are particularly susceptible to predatory animals such as

peregrine falcons and foxes. To survive, the ptarmigan has developed a unique camouflaging ability—it can change the color of its **plumage** (feathers) according to the season of the year.

In the winter, ptarmigans are almost completely white. The color of their plumage and ability to burrow into the snow makes them blend in perfectly with the rest of the landscape. In fact, it's even difficult to locate a whole group of ptarmigans walking across the snow.

In the spring and summer, the feathers of the ptarmigan change again. Males will have feathers that are gray or mottled brown in color, while females may have a variety of brown or black markings often touched with yellow. The colors blend with open patches of ground created by the melting snows.

During the autumn months, the ptarmigan's feathers become grayer in color, to help camouflage themselves among the rocks and dying plant life during this part of the year.

In addition to their changing colors, ptarmigans increase their chances of survival by crouching low on the ground at the first sign of a predator. By staying close to the ground their plumage blends in with the surrounding area. It is only when danger is very close that they fly away.

The whiteness of the ptarmigan's feathers in winter helps prevent too much heat loss from the bird's body.

SARGASSUM FISH

Sargassum fish are named after the Sargasso Sea—a large expanse of water in the Atlantic Ocean filled with enormous beds of seaweed. This 3-inch fish can be found in many of the warmer areas of the Atlantic Ocean.

Swimming Seaweed

How would you like to be able to walk through a yard or across a park and be almost invisible? What if your body looked like it had branches, leaves, twigs, and other growths sprouting all over it—so many green and brown extensions that you looked just like a bush?

The sargassum fish looks so much like a piece of seaweed that it is almost impossible to locate when hiding. One look at this fish and you would think that it was a plant with fins. In fact, that's exactly how it is supposed to look! It lives in brown and gold colored seaweed—the same colors as the fish.

Another unique feature of this fish is its body. It is covered with a variety of flaps and extensions—that all look like pieces of floating seaweed.

Its skin, too, is covered with tattered flaps and spiky edges growing in all directions. Spiny **protrusions** (things that stick out) grow on its head and along its belly.

Its coloration and distinctive body parts conceal the sargassum fish in the sargassum weeds and other marine plants where it lives. There it remains, very still, hiding from its

enemies—usually larger fish that frequently visit seaweed beds. In this way it avoids becoming a meal for something else!

Its shape and color also help it hide while waiting for smaller fish to swim nearby. Since the sargassum fish has several yellow growths sticking out from its body, other fish think these look like pieces of food drifting through the water. Smaller fish swim up to nibble at the sargassum fish's growths and GULP—they quickly become a tasty meal!

Fantastic Fact

Although the sargassum fish is relatively small it can eat other fish nearly as large as itself.

LIVING STONES

There are approximately 50 different species of living stone plants—all of which inhabit the dry, arid regions of South Africa.

Pretty Pretenders

Animals aren't the only organisms that use camouflage to survive in the wild. There is a group of special plants—known as living stones—that also uses camouflage to their advantage.

As you might imagine, life in the desert can be harsh and severe. Both plants and animals need to have special ways of surviving in this sometimes desolate environment. Living stones

form a solid stone-like body. Their shape makes them look like rocks with small cracks etched across their tops. To complete their camouflage, the leaves are "sprinkled" with small dots or flecks that resemble small bits of minerals. As a result, these plants appear to be just another group of desert rocks.

When the rains come, each of these plants blossoms with a single tiny flower—usually yellow or white—that pokes up between the tightly bunched leaves. The flowers do not last for very long and pollination, usually by wind or insects, occurs very rapidly.

Living stones have been able to adapt to a cruel and severe environment. Their size, shape, and color protect them from the long dry spells of the region as well as from desert animals looking for a quick meal.

are also known as stone-plants, stonefaces or pebble plants. They are able to survive because they have characteristics of something they are not—a group of pebbles or small stones.

Rainfall is infrequent in the desert, and plants that live in these regions of the world must be able to **germinate** (sprout), grow, flower, and produce seeds very rapidly.

Usually less than 1 inch tall,

these ground-hugging plants have paired leaves that are joined beneath the ground to

Each leaf has a collection of very tiny "windows" that permit light to reach inside the plant.

PROTECTING CLEVER CAMOUFLAGERS

The organisms in this book represent some of the most distinctive and unusual creatures on this planet. The ability of many animals and plants to camouflage themselves is how those organisms are able to survive in the places they live.

A group of living organisms of a single species living in a specific place is known as a population. Each population of animals shares some space with the populations of other animals. Clusters of animal populations living together are known as communities. All living things in a community are affected by all the other living things (as well as some non-living things such as water, soil, and rocks). The study of these relationships is called ecology.

For many animals, just living from day to day is a constant struggle. They must compete for food and they must avoid becoming a food source for other creatures, too. Often, that daily struggle for survival becomes complicated when humans enter the picture. When people build housing developments or pump dangerous toxins into the air with their cars or factories, then a nearby community of animals may be seriously affected. Some animals may not be able to hide because the plant life has been destroyed; other animals may not be able to find necessary food because the food has died as a result of being exposed to hazardous chemicals. The introduction of non-living things into an environment (for example, pollutants, highways, and buildings) has an effect on the survival and existence of all the living things in that environment.

Protecting and preserving wildlife and vegetation is an important concern around the world. What we put into an environment may have an impact on the survival of the many different communities that exist within that environment. In short, we affect how animals and plants are able to reproduce, grow, and live. That means that we must all work together to prevent or eliminate these environmentally harmful conditions.

Take some time to talk with your friends or classmates about ways in which humans affect the environment. Ask your school or local librarian for suggested books on environmental issues and problems.

Take a walk around your local neighborhood and note the various types of plants and animals that live there. What are some dangerous or harmful things in that environment that might affect the lives of those organisms? What can you and your friends do about those conditions? What you do now can have a positive affect on plants and animals in the years ahead.

Working with other people can make a difference in the survival of animal and plant populations. Find out as much as you can and share your knowledge with others. Your work is important to all living things.

Sea Dragon

Greenpeace
1436 U Street NW
Washington, D.C. 20009
(This organization deals with major issues affecting the oceans of the world. These issues may include whaling, pollution, and over-fishing.)

Izaak Walton League
1401 Wilson Boulevard
Level B
Arlington, VA 22209
(Concentrates on the cleanup and preservation of natural waterways such as streams and rivers.)

Nature Conservancy
1815 North Lynn Street
Arlington, VA 22209
(Purchases large pieces of land throughout the United States and around the world. This organization works to preserve those lands in their natural state.)

MORE CLEVER CAMOUFLAGERS

For many animals and some plants, survival is part of their day-to-day lives. For many organisms, their lives are very long simply because they have learned how to survive within their environments. For other organisms, their life span may be relatively short. Camouflage permits many organisms to find necessary food while avoiding becoming dinner for something else. Here are some other animals from around the world that have distinctive and unusual means of camouflage. You may want to learn more about them in your school or public library.

DECORATOR CRAB

The decorator crab covers its entire body with a variety of plants. Unsuspecting fish, thinking it's just another sea plant, get too close and become dinner for this crab.

WINDOWPANE BUTTERFLY

This South American insect has transparent wings—making it seem almost invisible to any predator passing by.

FROGMOUTH

This Australian bird rests on the trunks of giant jungle trees. As long as it doesn't move, it looks exactly like a dead tree trunk.

MALAYSIAN GRASSHOPPER

This Southeast Asian insect looks like a dead leaf. It can lie perfectly still on the forest floor for hours.

BRIMSTONE BUTTERFLY

This clever animal hibernates in the winter by hanging upside down in tree branches. Because its wings look like two dried-out yellow leaves hanging close together, hungry birds often pass it by.

TREE HOPPERS

These insects rest in groups on the stems of plants. They look exactly like thorns and escape detection from hungry enemies.

LEAF FISH

This South American fish lies on the bottom of streams and rivers. Its color and shape make it look exactly like a dead, fallen leaf.

NEW GUINEA BEETLE

The skin of this tropical beetle is covered with holes and crevices which hold tiny amounts of water. Small plants are able to grow in these holes and completely cover the beetles for camouflage.

KATYDID

The wings of this tree-dwelling insect have veins and other plant-like markings. With its bright green color, it looks exactly like a typical tree leaf.

FLOUNDER

This fish, which is usually brown, can change its color to match the colors of the ocean bottom.